JOINT INSTITUTE
FOR NUCLEAR RESEARCH

DUBNA

Survey

AKADEMIE-VERLAG · BERLIN
1958

Compiled by Anatoli Mikhailovich Ryzhov

Erschienen im Akademie-Verlag GmbH, Berlin W 8, Mohrenstraße 39
Lizenz-Nr. 202 · 100/532/58
Satz und Druck: Druckhaus „Maxim Gorki", Altenburg
Bestell- und Verlagsnummer: 5337 · ES 20 D 6
Printed in Germany

I. ESTABLISHMENT AND PURPOSES OF THE INSTITUTE

On the initiative of the USSR a conference of representatives of eleven countries — Albania, Bulgaria, Hungary, German Democratic Republic, Poland, China, Korea, Mongolia, Rumania, USSR and Czechoslovakia — assembled in Moscow on March 26, 1956. The representatives of the eleven countries recognised the necessity of joining scientific efforts in nuclear research and of founding the Joint Institute for Nuclear Research. There was elected a Management of the Institute, consisting of the Director of the Institute and two Vice-Directors. The Management was charged to prepare a Statute of the Joint Institute and an Instrument for its Personnel.

In September 1956 a Conference of the Plenipotentiary Representatives of the Governments of the Member States of the Institute took place, which approved the Statute and the Instrument for the Personnel. At this time Vietnam joined the Institute as the twelfth Member State.

The Soviet State handed over without compensation to the Joint Institute: 1. The Institute of Nuclear Problems of the USSR Academy of Sciences with a 680 MeV synchrocyclotron; 2. The Electrophysical Laboratory of the USSR Academy of Sciences with a 10 BeV synchrophasotron. These constructions became the material and technical basis of the Joint Institute.

As established by the Conference of Plenipotentiaries the purposes of the Institute are:

a) to make possible the carrying out of joint theoretical and experimental research in nuclear physics by scientists of the Member States;

b) to contribute to the development of nuclear physics in the Member States by exchange of experience and achievements in theoretical and experimental research;

c) to maintain contacts with national and international research institutions and other organizations concerned with a view of developing nuclear physics and finding new ways of using atomic energy for peaceful purposes;

d) to contribute to the all-round development of creative abilities of research workers in the Member States.

All activities of the Institute shall promote the use of atomic energy for peaceful purposes only and for the benefit of all mankind.

Results of the research work carried out by the Institute shall be published or reported to scientific conferences and sessions for general information. Reports on the investigations performed shall be sent to all Member States. The Member States of the Institute are the 12 above mentioned States that signed the Agreement on the founding of the Institute. The Statute allows for other states to be admitted to the Institute, if they desire to take part in the work of the Institute and accept the provisions of the Agreement on the foundation and its Statute, by a majority decision of the Member States.

All Members of the Institute have equal rights. In accordance with the Budget approved, each Member State pays within fixed terms financial contributions for the maintenance of the Institute and for new constructions. The largest contribution is being paid by the USSR — 47,25%, followed by China — 20%, Poland — 6,75%, German Democratic Republic — 6,74%, etc.

The Institute consists of the Management and Administration and of five Laboratories: Laboratory of Nuclear Problems, Laboratory of High Energy, Laboratory of Theoretical Physics, Laboratory of Neutron Physics, Laboratory of Nuclear Reactions. Besides, the Radiochemical Laboratory and the Plant of Physical Equipment are planned to be constructed.

The Joint Institute is governed by the Conference of the Plenipotentiary Representatives of the Governments of Member States which adopts the organization structure and Budget of the Institute. This Conference meets once a year. For the time between the Conferences all the practical problems are to be carried out by the Management of the Institute with the Director, to be elected for 3 years, and two Vice-Directors elected for 2 years.

The first Director of the Institute, being in charge today, is the Corresponding Member of the USSR Academy of Sciences, Professor D. I. Blokhintsev; two Vice-Directors are Professors V. Votruba (Czechoslovakia) and M. Danysz (Poland).

The Academic Council was organized in order to consider the plans for scientific work. The names of its members are well known in the scientific world:

D.I. Blokhintsev, H. Barwich, Chao Chun-chao, M. Danysz, V.P. Dzhelepov, E. Djakov, G. Hertz, H. Hulubei, L. Infeld, L. Jánossy, Kim Hyun Bong, Le-Van-Thiem, G. Niewodniczański, G. Nadjakov, V. Novacu, K. Novobatzky, Ning Hu, H. Pose, V. Petržilka, P. Pilika, C. Simàne, S. Szalay, N. Sodnom, A. Zoltan, S. Ťiteica, Ten Gyn, I. E. Tamm, V. I. Veksler, V. Votruba, Wang Kan-Chang.

The Director of the Institute is the Chairman of the Academic Council. Specialists of all the Member States are working in the Institute. Besides the collaborators from the USSR 125 research workers from the Member States are at present working here. This number increases systematically.

The Director of the Institute appoints an Administrative Manager to deputize for him in dealing with the administrative and economic activities of the Institute. At present V.N. Sergienko is holding this office.

Moscow Meeting of the Plenipotentiaries of Governments of Eleven States, March, 1956. The signing of the Agreement on the establishment of the Joint Institute for Nuclear Research

The Session of the Committee of the Plenipotentiaries of the Governments of States-Members of the Joint Institute for Nuclear Research

The Session of the Committee of the Plenipotentiaries of the Governments of States-Members of the Joint Institute for Nuclear Research. E. P. Slavsky, representative of the USSR, is taking the floor. Sitting are (from left to right): Prof. Kožešnik (Czechoslovakia), Prof. D. I. Blokhintsev, Corresponding Member of the USSR Academy of Sciences, Prof. M. Danysz (Poland), Prof. V. Votruba (Czechoslovakia)

The representatives of the German Democratic Republic, headed by Prof. K. Rambusch at the Session of the Committee of the Plenipotentiaries of the Governments of State-Members of the Joint Institute for Nuclear Research

Vice-Directors Prof. M. Danysz (Poland) and Prof. V. Votruba (Czechoslovakia) at the Session of the Academic Council

Chinese and Korean delegates at the Session of the Acedemic Council

Polish delegation at the Session of the Academic Council. From left to right Prof. G. Niewodniczanski, Academician L. Infeld, Prof. A. Zoltan

A group of scientists of the Joint Institute for Nuclear Research. From left to right Ya. A. Smorodinsky, D. I. Blokhintsev, I. Ulehla, V. P. Dzhelepov, Kim Hyun Bong, Wang Kang-Chang, N. Sodnom

Members of the Academic Council of the Joint Institute for Nuclear Research
Prof. H. Hulubei, Prof. S. Szalay, Prof. K. Novobátzki

II. THE LABORATORIES OF THE INSTITUTE

1. The Laboratory of Nuclear Problems

The Laboratory of Nuclear Problems was set up on the basis of the former Institute of Nuclear Problems of the USSR Academy of Sciences. The Director of this Laboratory is Prof. V.P. Dzhelepov.

The Academic Council of the Laboratory has the right to confer the degrees of Doctor or Candidate of Physico-Mathematical or Technical Sciences.

The following scientists are members of the Council:

V. P. Dzhelepov, V. P. Dmitrievsky, G. N. Flerov, I. M. Frank, E. Katz, M. S. Kozodaev, M. G. Mescheryakov, V. A. Petukhov, B. Pontecorvo, H. Pose, V. S. Siksin, J. A. Smorodinsky, V. I. Veksler, V. Votruba, Z. Wilhelmi.

The Director of the Laboratory is the Chairman of the Academic Council. The Laboratory of Nuclear Problems possesses a synchrocyclotron which produces protons of 680 MeV, deuterons of 420 MeV, and α-particles of 840 MeV. The main parts of this accelerator are:

1. An electromagnet, in the field of which the accelerated ions move;
2. A high frequency resonance system, producing an alternating electric field on the accelerating electrodes;
3. A vacuum accelerating chamber containing the accelerating dee and placed between the poles of the electromagnet;
4. An ion source, placed in the center of the vacuum chamber.

The electromagnet of the synchrocyclotron has the diameter of poles equal to 6 meters. It is made of steel. The field strength in the gap under working conditions is 16650 oersted.

The pole pieces which are at the same time the covers of the acceleration chamber, provide a monotonous decrease of the magnetic field from the centre to the edges necessary for focusing.

The accelerating electric field is produced by an oscillator, the resonance frequency of which is modulated in the necessary limits by means of a capacity variator.

The oscillating circuit of the system consists of the dee, a resonant line, the variator and is excited by a vacuum-tube oscillator.

The standing waves arising in the system produce the accelerating voltage between the dee and the grounded frame. The acceleration in the synchrocyclotron is cyclic, the duration of cycles depending on the parameters of the acceleration.

The vacuum chamber is a rectangular parallelepiped of thick sheet brass. The chamber volume is 33 m^3. The vacuum volume of the chamber and apparatus connected with it are limited by a metal surface of several hundred sq. meters and has some hundred different rubber gaskets.

To stabilize the motion of the accelerated beam, the accelerating chamber and the frequency variator are under a high vacuum of 3.10^{-6} mm Hg, which is provided by three powerful high speed diffusion pumps.

Extraction of the Beam from the Synchrocyclotron Chamber

The extraction of protons from the accelerator is performed by injecting the protons into a magnetic channel, which is formed by two iron plates, the form and the dimensions of which are specially chosen. For this purpose, an artificial perturbation of the magnetic field leading to increasing radial oscillations is being created on the last orbits of the accelerated beam. The protons which lost radial stability are peeled off from the equilibrium orbit, enter the magnetic channel, and leave the accelerating chamber. The intensity of the external proton beam at 15 m from the chamber is $1.5 \cdot 10^9$ protons/cm^2 sec. The synchrocyclotron, accelerating protons to 680 MeV, is also a powerful source of 600 MeV neutrons, charged and neutral π-mesons of 400 MeV, 300 MeV μ-mesons and 640 MeV polarized protons.

To obtain the above mentioned particles the accelerated proton beam hits an internal target (usually beryllium).

The particles, produced as a result of collisions, emerge from the target under a certain angle to the proton beam orbit and are ejected into the atmosphere through special dural windows.

To provide better conditions for experiments with protons, neutrons, mesons and polarized protons a laboratory was built, which is separated from the accelerator by a concrete wall and ceiling, protecting research apparatus from scattered radiations.

In the shielding wall there are channels, with steel collimators, which allow to separate narrow proton, neutron and meson beams.

At present 17 beams of different particles are extracted from the chamber. About 50 different experimental set-ups are aligned in the way of these particles. Under some working conditions there is a possibility to carry out investigations simultaneously with 5 beams using at the same time about 10 experimental set-ups.

During the operation of the synchrocyclotron the personnel is not present in the main hall due to radiation danger. The remote control and regulation are realized from the other building. The accelerator works 140 hours per week. The operation of the accelerator is supervized by the chief engineer B. I. Zamolodchikov.

The main directions of scientific research with the synchrocyclotron are investigations with high energy particles of the problems of nuclear forces and of the properties, and structure of elementary particles.

During the operation of the synchrocyclotron (starting from December, 1949) a great number of scientific investigations has been performed on the study of nuclear interactions of high energy particles (nucleons, π-mesons, μ-mesons) and of the properties of these particles. The results were published in Soviet and foreign journals and were discussed at Soviet and international conferences. Various modern methods of investigations are employed in the experiments: scintillation and Čerenkov counters with photomultipliers and with complicated electronics; diffusion, bubble and cloud chambers of different dimensions in magnetic field; multichannel magnetic spectrometers, photographic emulsions, radiochemical methods, etc.

The Laboratory has a designing office and experimental mechanical shops allowing to make complicated unique devices. A considerable group of scientists headed by Prof. V.P.Dzhelepov works in this Laboratory. The Laboratory consists of several sectors, headed by the Corresponding Members of the Academy M. G. Mescheryakov, B. Pontecorvo, Professors H. Pose, T. Tanasescu, and Candidates of Physico-Mathematical Sciences V. P. Dmitrievsky, R. M. Sulyaev and others.

Synchrocyclotron building of the Laboratory of Nuclear Problems

Synchrocyclotron of the Laboratory of Nuclear Problems

Laboratory of Nuclear Problems. Bubble chamber with liquid hydrogen

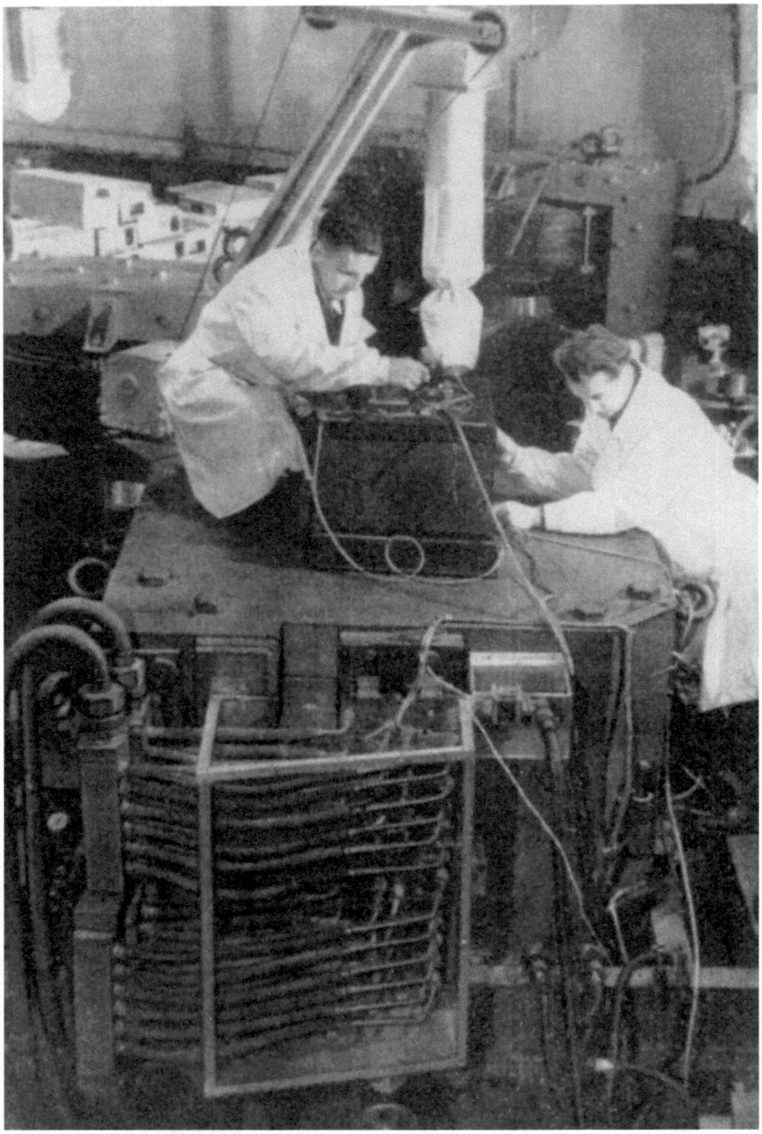

Laboratory of Nuclear Problems. A cloud chamber in the magnetic field

Laboratory of Nuclear Problems.
The arrangement with the diffusion chamber in the magnetic field

Laboratory of Nuclear Problems. Magnetic quadrupole lens

Laboratory of Nuclear Problems. Detecting arrangement for work with a pion beam

Laboratory of Nuclear Problems. Bubble chamber with propane

Laboratory of Nuclear Problems.
Apparatus for the experiments with a polarized proton beam

2. The Laboratory of High Energies

Academician V. I. Veksler is in charge of this Laboratory.

The Laboratory has its Academic Council, having the right to confer degrees of Doctor or Candidate of Physico-Mathematical or Technical Sciences.

The following scientists are members of this Council:

V. I. Veksler, Wang Kan-Chang, J. Werle, V. P. Dzhelepov, M. Danysz, A. A. Kolomensky, M. G. Mescheryakov, M. A. Markov, A. L. Mintz, B. Pontecorvo, M. I. Podgoretsky, V. A. Petukhov, I. E. Tamm, K. D. Tolstov, T. Tanasescu, I. V. Chuvilo.

The Director of the Laboratory is the Chairman of the Academic Council of the Laboratory.

The Laboratory of High Energies is equipped with a 10 BeV synchrophasotron which on April 16, 1957 produced first protons at the design energy of 10 BeV. The synchrophasotron is a ring type accelerator of 56 meters diameter. The accelerator magnet consists of 4 quadrants, made of special Ш-shaped steel. The accelerator magnet is supplied with current from rectifyiers and the latter — from 4 generators of the total capacity 140 000 kw per pulse.

The doughnut vacuum chamber is placed between the magnet poles inside of which the accelerated particles move. The vacuum chamber consists of two parts: the high vacuum part and the vacuum chamber for preliminary rarefaction around the high vacuum chamber.

The pumping of the high vacuum chamber is performed by 56 vacuum pumps. The particle injection is realized by a 9 MeV linear accelerator. One interval between the magnetic quadrants is used for the particle injection. There are accelerating electrodes in two other intervals. The fourth interval is used for the beam ejection. The accelerator is operated by remote control.

Particle Beams

At present the internal beam is in operation. It is supposed to obtain a proton beam in the measuring pavilion which is separated from the proton synchrotron (synchrophasotron) hall by a concrete wall 8 m thick. Besides that, it is supposed to obtain a π-meson and neutron beam in the measuring pavilion. K-mesons with $p = 420$ MeV/c will be ejected through a window on the outer side of the magnet ring. K-mesons and π^+-mesons will be ejected also inside the ring.

The beam focusing will be carried out by lenses with strong focusing. The most interesting parameters of the accelerator are given below:

1. Maximum proton energy — 10^{10} eV
2. Magnet weight — 36 000 t
3. Orbit diameter — 56 m
4. Maximum field — 13 000 oersted
5. Time of the acceleration cycle — 3,2 sec
6. Energy increment per revolution — 2,3 KeV
7. The lenght of the rectilinear interval — 8 m
8. Number of revolutions which particles make at acceleration to maximum energy — $4{,}5 \cdot 10^6$

The scientific work of the Laboratory is concentrated in the following directions:

a) Investigation of the synchrophasotron itself and further development of new methods of charged particle acceleration. Professors V. I. Veksler, V. A. Petukhov and others work in this field;

b) Experimental investigations with protons and secondary particles from the 10 BeV machine, namely: interactions of „elementary" particles and antiparticles with nucleons and nuclei, investigation of their properties and the search for new particles.

This work is carried out by a group of scientists, under the leadership of the Deputy Director of the Laboratory, Candidate of Physico-Mathematical Sciences I. V. Chuvilo.

A group under Prof. Wang Kan-Chang is engaged in the development of bubble chamber techniques.

The Laboratory is working now on the preparation of the following techniques:

a) scintillation and Čerenkov counters of different types with electronic equipment;

b) large bubble chambers, a diffusion chambers of about 2 m length, cloud chambers;

c) emulsion stacks and plates.

The main direction of research is the physics of elementary particles. The study of elementary particle properties and their interactions, investigation of their production characteristics, interaction and decay of different particles: nucleons, mesons, hyperons and anti-particles and search for new particles.

Academician V. I. Veksler is reporting to the Academic Council

Main control panel of the synchrophasotron of the High Energies Laboratory

A part of the synchrophasotron magnet of the High Energies Laboratory. At the front one of the diffusion pumps

Arrangement for beam injection into the synchrophasotron chamber

Main machine hall of the power building of the High Energies Laboratory.
The generators supplying the windings of the synchrophasotron magnet

3. *The Laboratory of Theoretical Physics*

The Joint Institute comprises a Laboratory of Theoretical Physics. Before setting up the Joint Institute, the Institute of Nuclear Problems and the Electrophysical Laboratory had theoretical sectors. On the basis of these sectors the Laboratory of Theoretical Physics was founded. At present, the Laboratory has 8 sectors with 95 collaborators. The Laboratory possesses a computing sector equipped with electronic computers. The building of the Laboratory is under construction and will be ready by the beginning of 1959, and therefore nowadays this Laboratory is placed in the buildings of the Laboratory of Nuclear Problems and Laboratory of High Energies.

The Director of the Laboratory is Academician N. N. Bogolyubov. The sectors are headed by: Corresponding Member of the USSR Academy of Sciences M. A. Markov, Professors J. A. Smorodinsky, Ning Hu (China), S. V. Fomin; Prof. J. Rzewuski (Poland) who now returned to Poland. A large group of Candidates of Physico-Mathematical Sciences is working in the Laboratory: A. A. Logunov, V. G. Soloviev, D. V. Shirkov, L. I. Lapidus, N. P. Klepikov, B. V. Medvedev, V. N. Ogievetsky, V. S. Barashenkov and others.

The Theoretical Laboratory has its Academic Council with the right to confer degrees of Doctor or Candidate of Physico-Mathematical or Technical Sciences.

The following scientists are members of this Council:

N. N. Bogolyubov, D. I. Blokhintsev, V. Votruba, J. Rzewuski, M. A. Markov, G. Marx, V. I. Ogievetsky, J. Plebansky, J. A. Smoordinsky, S. V. Fomin, G. Heber, Ch. Christov, Ning Hu, D. V. Shirkov, S. Ţiţeica.

The Director of the Laboratory is the Chairman of the Academic Council.

The Laboratory is engaged on the following problems: quantum field theory, theory of elementary particles and phenomenological theory of nuclear reactions etc.

Design of the building of the Theoretical Physics Laboratory.
Its construction will be completed in 1959

Theoretical Physics Laboratory. The computer "Ural"

4. The Laboratory of Neutron Physics

This Laboratory is in the state of construction. The building of the pulsed reactor has been started and it will be finished in the second half of 1958. The mounting of the equipment has begun; at the end of 1958 the fast pulsed reactor should start its work.

The Director of the Laboratory is Corresponding Member of the USSR Academy of Sciences Professor I. M. Frank.

The Laboratory has a small group of collaborators working on the development and preparation of experimental arrangements, design etc.

The plan of the Laboratory includes problems of neutron spectroscopy, nuclear reactions induced by neutrons and other problems.

5. The Laboratory of Nuclear Reactions

During the last years a great deal of attention in the field of nuclear studies is directed to the investigation of nuclear interactions by multiply ionized atoms.

These investigations give not only a great number of new isotopes but are promising methods of investigating new transmendeleevian elements.

Taking this into consideration the first Session of the Academic Council of the Institute in 1956 decided to construct a cyclic accelerator especially for the acceleration of multiply charged ions.

The successful carrying out of the design and construction of this cyclotron and the obtaining of some new data in investigations conducted by the method of multiply charged ions showed the necessity of creating a special laboratory which can develop the investigations of nuclear reactions induced by multiply charged ions.

Therefore in November 1957, the third Session of the Academic Council has decided to set up the Laboratory of Nuclear Reactions.

The Director of the Laboratory is the Corresponding Member of the USSR Academy of Sciences G. N. Flerov.

The Laboratory will be equipped with a cyclic accelerator.

The construction of the magnet for this cyclotron is now finished, and from 1958 the mounting was started. Other equipment is being delivered during the first half of 1958.

The Laboratory will be placed in a building now under construction. Besides the main hall with the accelerator, the Laboratory will have some other buildings where physical and radiochemical laboratories will be sited.

Design of the main building of the Neutron Physics Laboratory. It will be built in 1958

Design of the building of the Neutron Physics Laboratory. Main facade

Design of the buildings of the Nuclear Reactions Laboratory. The construction will be completed by the end of 1958

III. INTERNATIONAL CONTACTS OF THE JOINT INSTITUTE FOR NUCLEAR RESEARCH

One of the purposes of the Joint Institute for Nuclear Research is „to maintain contacts with national and international research and other organizations concerned, with a view to developing nuclear physics and finding new ways of using atomic energy for peaceful purposes". (Article 4 of the Statute.)

The Institute develops its scientific international contacts by participating in the international conferences, by holding scientific conferences and meetings at the Institute, by visits of its scientists to other research organizations, by inviting to the Institute the scientists from other countries, as well as by exchange of scientific information.

The scientists of the Joint Institute took part in some international conferences on nuclear physics, including the International Moscow Conference on High Energy Nuclear Physics (May 1956), CERN Symposium, held in Geneva (June 1957), VI Rochester Conference (April 1957), International Stanford Conference on Nuclear Sizes (December 1957), International Conference on High Energy Nuclear Physics in Geneva (July 1958), as well as in other Conferences and Meetings which took place both in Member States of the Institute and in other countries.

For a short period of time some international meetings on working level with discussion of principal directions of scientific investigations were held at the Joint Institute, including the meeting on photoemulsion methods and the meeting on different kinds of chambers, the meetings on neutron physics, radiochemistry and nuclear spectroscopy.

The scientists of the Joint Institute are visiting the Member States to deliver lectures, to take part in seminars, to exchange experience in research work. Among them were: Director of the Joint Institute Professor D. I. Blokhintsev, Academician N. N. Bogolyubov, Professors V. Votruba, M. Danysz, V. P. Dzhelepov, J. A. Smorodinsky, Ning Hu, research workers M. I. Podgoretsky, I. M. Gramenitsky, S. I. Lyubomilov and others.

The scientists from the countries which are not members of the Institute take great interest in its work. Such outstanding scientists as Professors F. Joliot-Curie † (France), C. F. Powell (England), A. Bohr (Denmark), W. K. H. Panofsky, E. Segrè, L. W. Alvarez, V. F. Weisskopf (the USA), H. J. Bhabha (India), Z. Koba, H. Yukawa (Japan) and others visited the Joint Institute. Some of them delivered lectures and took part in the work of scientific seminars.

Sometimes scientists from the countries which are not members of the Institute stay at the Institute for a long period of time. Thus, in 1957, a French physicist theoretician, George Lochaque, worked here for a year. This year another French research worker Jeanne Laberrigue-Frolow came to the Institute with a six month visit.

In accordance with Article 4 Chapter II of the Statute the results of the research carried out at the Instituts are published in scientific magazines, reported to scientific conferences and, as a rule, are sent in the form of preprints to research centres of many countries as mutual exchange of scientific information.

At present the Institute has established the exchange of scientific information with dozens of leading research laboratories and institutes of the world, as well as with many scientists working in the field of high energy physics.

Developing international scientific cooperation the Joint Institute for Nuclear. Research in accordance with Article 4 Chapter II of the Statute shall: ,,promote the use of atomic energy only for peaceful purposes to the benefit of all mankind.“

President of the Czechoslovak Republic A. Zapototsky † (in the centre) at the Laboratory of Nuclear Problems

Visit of the Government delegation of the People's Republic of Bulgaria to the High Energies Laboratory. From left to right: V. A. Petukhov, Vice-Director of the High Energies Laboratory, L. Gerasimov, the Ambassador of the People's Republic of Bulgaria in the USSR, A. Yugov, the Chairman of the Council of Ministers of the People's Republic of Bulgaria, D. I. Blokhintsev, the Director of the Joint Institute for Nuclear Research

Visit of Prof. F. Joliot-Curie † to the Joint Institute for Nuclear Research
In the centre Prof. F. Joliot-Curie and Director of the Institute Prof. D. I. Blokhintsev. To the left the scientific worker of the Joint Institute for Nuclear Research Jeanne Laberrigue

Visit of Gamal Abdel Nasser,
President of the United Arab Republic, to the Joint Institute for Nuclear Research

Visit of the Secretary General of the United Nations Mr. D. Hammarskjoeld to the Joint Institute for Nuclear Research (the second from the right)

Professor Dirac at the Joint Institute for Nuclear Research, accompanied by Director of the Institute Prof. D. I. Blokhintsev, Vice-Director Prof. M. Danysz, Corresponding Member of the USSR Academy of Sciences M. G. Mescheryakov, Academician N. N. Bogolyubov and Prof. Y. A. Smorodinsky

Mr. Perrin in the High Energies Laboratory. On the right D. V. Efremov, Deputy-Head of the Atomic Energy Utilization Board under the Council of Ministers of the USSR

A group of American scientists visited the Joint Institute for Nuclear Research.

From left to right: Prof. W. Panofsky (USA), Prof. L. Smith (USA), Prof. E. Segrè (USA), Prof. O. Chamberlain (USA) among the scientists of the Joint Institute for Nuclear Research: Prof. V. I. Goldansky, Corresponding Member of the USSR Academy of Sciences G. N. Flerov, Corresponding Member of the USSR Academy of Sciences M. G. Mescheryakov, Prof. M. S. Kozodajev, Prof. V. P. Dzhelepov

The Corresponding Member of the USSR Academy of Sciences M. G. Mescheryakov is demonstrating the installations of the Joint Institute to a group of American scientists

General view of the synchrophasotron electromagnet of the High Energies Laboratory

Bei Fragen zur Produktsicherheit wenden Sie sich bitte an:
If you have any questions regarding product safety,
please contact:

Walter de Gruyter GmbH
Genthiner Straße 13
10785 Berlin
productsafety@degruyterbrill.com